Jaqueline Ap. Gonçalves Soares
Isane Vera Karsburg

Mutagenesis of aqueous solutions using plant bioindicators

AF665184

Jaqueline Ap. Gonçalves Soares
Isane Vera Karsburg

Mutagenesis of aqueous solutions using plant bioindicators

Cytotoxicity and genotoxicity

ScienciaScripts

Imprint
Any brand names and product names mentioned in this book are subject to trademark, brand or patent protection and are trademarks or registered trademarks of their respective holders. The use of brand names, product names, common names, trade names, product descriptions etc. even without a particular marking in this work is in no way to be construed to mean that such names may be regarded as unrestricted in respect of trademark and brand protection legislation and could thus be used by anyone.

Cover image: www.ingimage.com

This book is a translation from the original published under ISBN 978-3-330-76059-2.

Publisher:
Sciencia Scripts
is a trademark of
Dodo Books Indian Ocean Ltd. and OmniScriptum S.R.L publishing group

120 High Road, East Finchley, London, N2 9ED, United Kingdom
Str. Armeneasca 28/1, office 1, Chisinau MD-2012, Republic of Moldova, Europe
Printed at: see last page
ISBN: 978-620-8-34085-8

Copyright © Jaqueline Ap. Gonçalves Soares, Isane Vera Karsburg
Copyright © 2024 Dodo Books Indian Ocean Ltd. and OmniScriptum S.R.L publishing group

CONTENTS

CHAPTER 1	2
CHAPTER 2	6
CHAPTER 3	23
CHAPTER 4	28
CHAPTER 5	31
CHAPTER 6	36
CHAPTER 7	40
CHAPTER 8	41

CHAPTER 1

CYTOTOXICITY AND GENOTOXICITY

Living beings are exposed to the actions of countless toxic agents that can cause health problems. Medicines based on medicinal plants have been widely used, but most of the time they are based on popular knowledge and lack the authorisation and evidence to confirm their efficacy.

Before these medicines can be used, experiments need to be carried out to identify which substances they are and what functions they fulfil. The science that studies toxic substances and their action on living organisms is called Toxicology.

Research has been carried out to show that environmental mutagens can occur spontaneously. Most of the time they are induced by physical, chemical or biological agents, which can cause risks to human health.

Of the physical, chemical or biological substances, chemical substances offer the greatest risk and alterations to genetic material, as they affect the phases of cell division such as interphase, cytokinesis and the cell cycle, leading cells to apoptosis. The ability of a toxic agent to compromise the DNA molecule is called genotoxicity and is related to pollution.

Genotoxic compounds are found in all ecosystems and these chemical substances are cumulative in living beings and can have deleterious effects on exposed organisms.

Plants considered medicinal contain substances present in the stem, leaf, root, flower and fruit that can be used for therapeutic purposes and in alternative medicine.

The use of these substances for medicinal purposes is being increasingly valued, which is why more knowledge is needed about their therapeutic properties and their toxic potential for organisms.

According to the World Health Organisation, around 60-80% of the world's population has no access to traditional medicine, relying fundamentally on plants and their therapeutic substances for health care. Despite the fact that the world's plant genetic diversity is significant, less than 20% of species are scientifically studied (DA SILVA et al., 2015). The study of medicinal plants has been growing due to their easy acceptability, availability and low cost (GADELHA et al., 2013).

Some medicinal plants in popular use can have toxic properties, so the misconception that plants are natural remedies and therefore free of risks and side effects must be revalidated. Obeying the prescribed dosages and taking care to accurately identify the material used can prevent a number of accidents (LORENZI et al., 2000). The undesirable effects of medicinal plants are often lower when compared to well-defined synthetic drugs (SANTOS et al., 2016).

According to TRAESEL et al. (2015) and BRITO (2008), scientific studies are mandatory for the appropriate use of their active compounds. Biomarkers are useful and can be defined as indicator systems that generally include subsystems of a complete organism, used to identify a specific target (SANTOS et al., 2011).

Bioassays carried out on plants have been considered more sensitive when compared to animals (FACHINETTO et al., 2007) and have demonstrated the action of these organisms in monitoring mutagenicity caused by environmental pollutants (FERNANDES et al., 2007; MASCHIO, 2009).

One of the ways to identify these possible harmful effects is to assess the cytotoxicity of substances by means of alterations in the process of cell division in the test organism and by the incidence of chromosomal mutations, such as the loss or breakage of chromosomes or the development of micronuclei (SOUZA et al., 2005).

The evaluation of genotoxicity causes the occurrence of cells with anomalies and increased concentrations, increased cell division and chromosomal aberrations. Genotoxic effects can cause DNA breaks, resulting in the loss of genetic material and mutations that render cells unviable or lead to carcinogenic processes (CARITÁ et al., 2008).

Cytogenetic studies confirm the possible chromosomal alterations in plants due to the presence of mutagenic agents (MENDES, 2008), inhibition of the cell cycle, interruption in metaphases, inference of numerical and structural chromosomal alterations and exchanges between sister chromatids, among others (FACHINETTO et al., 2007).

Biological toxicity and genotoxicity tests provide a degree of proximity when comparing probable toxic effects and are also indispensable for assessing the reactions of living organisms to environmental pollution (ANDRADE, 2012).

The tests can be considered a simple, fast, sensitive and low-cost method. In order to obtain significant results, we used some plant species as bioindicators. The species that showed the best results in our research was Allium cepa L. (onion), followed by Lactuca sativa (lettuce) and Pisum sativum (pea).

1.1. Cytotoxicity and genotoxicity using the Allium cepa bioindicator

The onion (Allium cepa L.) is grown practically all over the world and is the third most economically important vegetable in Brazil. In older taxonomic systems, the species belonged to the Liliaceae family and the Allioideae subfamily, but more recent taxonomists include it in the Aliaceae family. The term refers to the bulb, which is made up of scaly, layered leaves. The basic number of chromosomes in its cells is 2n = 16.

The genus Allium is found from the Arctic Circle to the European continent, Asia, North America and Africa. However, in contrast to most cultivated plants, the exact centre of origin of the A. cepa species still remains obscure. In Brazil, the onion is grown in several Brazilian states, such as Santa Catarina, Rio Grande do Sul, São Paulo, Minas Gerais and Paraná, for example.

It is believed that onion consumption helps prevent certain diseases, which characterises it as a functional food. The genotoxic potential of most medicinal plants that are popularly used has not yet been studied (SILVA et al., 2016).

The A. cepa test is validated by the International Programme on Chemical Safety (IPCS, WHO), and the United Nations Environmental Programme (UNEP) is efficient for in situ monitoring of the genotoxicity of environmental substances (FACHINETTO et al., 2007; ALVES, 2013). Among the species, A. cepa is the one most recommended as a standard test material by the Royal Swedidsh Academy of

Science (FISKESJO, 1993) and the Gene Tox Programme (MASCHIO, 2009).

In order to assess the damage that mutagenic agents can cause, it is necessary for the sample to be in constant mitotic division, so that the toxic effects and changes added over the course of a cell cycle can be identified, and the A. cepa test has been widely used for this purpose (SILVA et al., 2012; OLIVEIRA, 2014).

The mitotic index is an indicator of adequate cell propagation (GADANO et al., 2002), which can be measured by the A. cepa plant test system.

In the A. cepa test, the onion roots are in direct contact with the substance tested, allowing different concentrations to be evaluated. The effects of medicinal plant infusions on the cell cycle of A. cepa have been reported by some authors (LEITE et al., 2016; LIMA et al. 2016), who have verified the effects caused by various plant extracts, with a view to detecting genotoxicity (FACHINETTO et al., 2007; 2009). These studies showed that the main effects that occur are mutagenesis, such as increased and decreased cell proliferation of root tips treated with different medicinal plant species.

CHAPTER 2

exposure of allium cepa to infusions, juices and plant macerates

2.1. Morinda citrifolia (Linn) in natura

Morinda citrifolia (Linn) belongs to the evergreen Rubiaceae family and is small to medium-sized. M. citrifolia has popular names, depending on the region of each country it can be called noni, nono, nonu, ba ji tian, cheese fruit. In most regions, parts of the plant are used for medicinal purposes, such as the fruit, leaves, bark, trunk and roots (WANG et al., 2002).

One of the active biological principles of M. citrifolia is an alkaloid known as xeronine, which is useful in medicine, food and industrial fields. The composition, characterisation, mode of action and usefulness of this alkaloid can be isolated from a large number of natural substances using certain techniques and precautions (HEINICKE, 1985).

The most commonly used part of the plant is the fruit, which has applications such as: antibacterial, analgesic, anticongestive, antioxidant, expectorant, anti-inflammatory, astringent, emollient, emmenagogue, hypotensive, blood purifier, immunostimulant and tonic. There are also reports of its anti-cancer action.

The aim of this study was to assess the behaviour of A. cepa cells in different solutions of M. citrifolia and with different exposure times.

Negative and positive controls and a 4x4 factorial with concentrations of 5, 10, 15 and 20 grams exposed at 24, 48, 72 and 96 hours were used for each test system. The collected rootlets were washed and fixed in methanol: acetic acid (3:1) and stored under refrigeration at -4 °C for later use. 4000 cells were analysed per treatment.

The slides were prepared using the methodology of GUERRA et al. (2001). The mitotic index (MI) was obtained using the methodologies of OLIVEIRA et al. (1996) and PIRES et al. (2001).

The averages were analysed by variance and the Skott-Knott test was applied to significant variations at 5% probability (FERREIRA, 2011).

Results obtained at different concentrations and different exposure times for M. citrifolia showed MI for the bioindicator A. cepa of 5.34% and 0.78% for the negative and positive control, respectively. The MI of the negative control was higher than that of the concentrations evaluated. It was noted that the highest MI was 3.68% in 5 g L^{-1} and 4.68% in 10 g L^{-1} exposed over 24 hours (Table 1).

For the treatments using 20 g L^{-1} of the fruit, it can be seen that the MI was high when compared to the other concentrations, especially during the 72-hour exposure time. These chromosomal aberrations or abnormalities are recognised as important consequences of the genotoxic actions of chemical agents (NATARAJAN, 2002 apud OBE et al., 2004), to which many organisms, including humans, are exposed (LEME et al., 2008).

Table 1. Mitotic index of A. cepa obtained in different concentrations of M. citrifolia extract.

Treatment	Time (h)	CN (%)	IM (%)	CA (%)
C +	24	5,15	5,34a	0,98
C -	72	0,78	0,78b	0,13
5 g L^{-1}	24	3,65	3,65b	0,00
10 g L^{-1}		4,68	4,68b	0,00
15 g L^{-1}		2,33	2,33b	0,20
20 g L^{-1}		2,53	2,53b	0,18
5 g L^{-1}	48	1,98	1,98b	0,05
10 g L-1		2.50	2,50b	0,08
15 g L^{-1}		1,58	1,58b	0,20
20 g L^{-1}		3,33	3,34b	0,55
5 g L-1	72	1,85	1,85b	0,00
10 g L^{-1}		1,95	1,95b	0,13
15 g L^{-1}		2,65	2,66b	0,30
20 g L^{-1}		3,50	3,51b	0,15
5 g L^{-1}	96	2,63	2,63b	0,00
10 g L^{-1}		3,25	3,25b	0,00
15 g L^{-1}		2.28	2,28b	0,03
20 g L^{-1}		2,48	2,48b	0,28

CV (%)	-	-	3,76	-

Averages followed by different letters in the same column differ at the 5% probability level using the Skott-Knott test.
Abbreviations: AC = abnormal cells; NC = normal cells; CV = coefficient of variation; C + = positive control; C - = negative control; MI = mitotic index.

The most common chromosomal alterations observed in this study were isolated chromosomes in different mitotic phases in A. cepa roots and chromosome bridges in telophase and micronuclei in interphase (Figure 1).

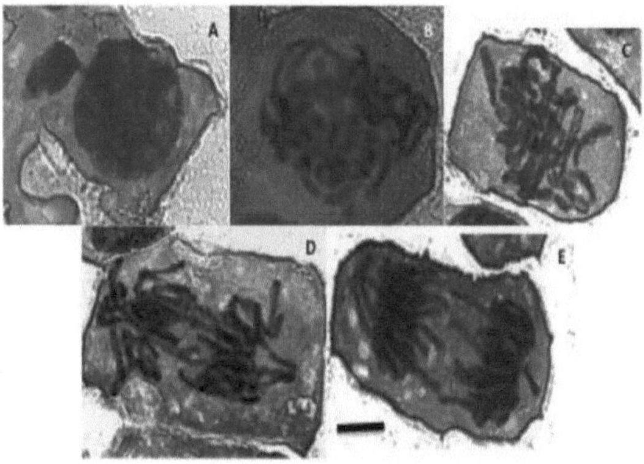

Figure 1. Mitotic cells of A. cepa with mutations. **A.** Interphase with micronucleus; **B.** Prophase with loop-shaped chromosome; **C.** Metaphase with isolated chromosome; **D** and **E.** Telophase with bridge and isolated chromosome. Bar = 5 ^m.

The A. cepa test system proved to be sensitive to exposure to the aqueous extract of M. citrifolia, corroborating the data of MULLER (2007). In conclusion, the analyses of the data obtained from the aqueous extract of M. citrifolia against the A. cepa biomarkers with the negative and positive controls show that M. citrifolia has a cytotoxic effect, as it inhibited cell division as observed in the significant decrease in the mitotic index.

2.2. Different concentrations of pyroligneous liquor from Tectona grandis Linn. F.

The species Tectona grandis L. F. (teak) is a tropical, arboreal, deciduous species belonging to the Lamiaceae family (LORENZI, 2002). It originated in the tropical forests of Southeast Asia and was introduced to Brazil, where the planted area has been growing in recent years, especially in the Mato Grosso region.

A plant species that has also been used to produce charcoal with a high calorific value (LORENZI et al., 2003), some plants produce compounds from the secondary metabolism that are known as allelopathic.

In the process of carbonising wood, the main product obtained is charcoal. The vast majority of carbonisations are carried out without recovering the gases. Condensing and recovering these volatile gases produces pyroligneous extract as well as charcoal.

The diversity of uses for pyroligneous liquor in a wide range of plants, from biofertiliser to fungicide and bactericide, has led to attention being drawn to its multiple uses without knowledge of the appropriate doses and the implications for human contact with these substances, as well as the care that must be taken to handle this product properly.

Because pyroligneous liquor is made up of more than 20 different substances which together can be toxic to animals and plants.

However, the study of cytogenotoxicity helps with preventive measures regarding the concentrations that can be used in the most appropriate way without harming the environment. The aim of this analysis was to assess the behaviour of Allium cepa mitotic cells in different concentrations of Tectona grande pyroligneous liquor.

The following concentrations of pyroligneous liquor were analysed: 10, 20, 30, 40 and 50 LIL L^{-1} , as well as the negative control, without the presence of pyroligneous liquor at 24, 48, 72 and 96 hours for each test system. The experiment had a 6 x 10 x 20 factorial design. The collected rootlets were washed and fixed in methanol:acetic acid (3:1) and stored under refrigeration at -4 °C for later use.

250 cells were analysed per treatment. The slides were prepared using the methodology of GUERRA et al. (2001). The mitotic index (MI) was obtained using the methodologies of OLIVEIRA et al. (1996) and PIRES et al. (2001).

The means were subjected to analysis of variance and, for significant variations, the Tukey test was applied at 5% probability (FERREIRA, 2011). The results obtained in the cytogenotoxicity assessments were analysed with 5,000 cells per treatment. In mitotic division, the prophase stage showed variations in cell distribution ranging from 2.38% to 34.86%; in the other phases of mitosis, the variations were more homogeneous. With regard to the irregularities observed in the different phases of mitosis, the cells with alterations ranged from 0.46% to 2.20% (Table 2).

Table 2. Evaluation of the cytogenotoxic effect on A. cepa cells of different concentrations of T. grandis pyroligneous liquor at different exposure times.

Treatment	Time (h)	CN (%)	IM (%)	CA (%)
C -	-	6,32	5,30b	0,70a
10 µL. L $^{-1}$	24	24,06	24,60a	0,46a
	48	25,16	12,40b	1,06a
	72	10.22	7,90b	0,62a
	96	20.22	20,20a	0,54a
20 µL L $^{-1}$	24	20,04	20,40a	0,94a
	48	8,88	8,80b	0,58a
	72	5,98	5,80b	0,66a
20 µL L $^{-1}$	96	12,72	12,70b	0,60a
30 µL L-1	24	4,82	4,80b	0,68a
	48	5,46	5,40b	0,76a
	72	3,70	3,70b	1,38a
	96	9,88	9,30b	1,10a
40 µL L $^{-1}$	24	10,48	10,40b	1,12a
	48	13,92	13,90b	1,76a
	72	22,70	20,20a	1,38a
	96	22,02	22,80a	1,14a
50 µL L $^{-1}$	24	28,58	28,50a	1,02a
	48	49,02	25,00a	1,86a
	72	35,72	35,70a	2,20a
	96	29,02	29,10a	1,72a
CV (%)	-	71,64	23,40	12,4

Averages followed by different letters in the same column differ at the 5% probability level using the Tukey test.
Abbreviations: AC = abnormal cells; NC = normal cells; CV = coefficient of variation; C + = positive control;

C - = negative control; MI = mitotic index.

Chromosomal alterations (CA) are recognised as important consequences of the genotoxic actions of chemical agents and can be observed in studies with onions: chromosomal irregularities, chromosomal bridges, lost chromosomes (DAMASIO et al., 2016).

The mitotic index (MI) of the negative control was 5.30 per cent. The treatments 20 ML L^{-1} for 72 hours and 30 ML L^{-1} for 48 hours, different concentrations and time exposure, the meristems of A. cepa showed mitotic indices close to the control treatment. In the 30 ML L^{-1} treatments for 72 and 24 hours, the mitotic indices were lower than the MI obtained in the negative control. In the other treatments, mitotic indices were higher than in the control Table 2.

The differences in MI observed were associated with the concentration and exposure time, with the highest indices being found with the highest concentrations. In the 50 μL L^{-1} treatment for 72 and 96 hours, a
IM was the highest, with the two highest mitotic percentages for cells in prophase, and among the lowest for cells in metaphase, anaphase and telophase, with the treatment of 50 ML L^{-1} for 72 hours also showing the highest number of cells with abnormalities. The treatment with the lowest MI was 30 ML L^{-1} for 72 hours (Table 2). According to FERNANDES et al. (2016), all the treatments analysed for the aqueous extract of Apiaceae on A. cepa showed no significant differences in MI compared to the negative control.

The structural alterations found in the cells were: micronuclei in intraphase cells (Figure 2B), prophase with an isolated chromatin fragment (Figure 2D), metaphase with an isolated chromosome (Figure 2F), anaphase with isolated chromosomes (Figures 2H-I) and telophase with an isolated chromosome (Figure 2L) and the formation of chromatin bridges (Figure 2M).

Thus, the presence of micronuclei in somatic cells is indicative of chromosome breaks (clastogenesis) and/or mitotic spindle disorders (aneugenesis) (NATARAJAN, 2002).

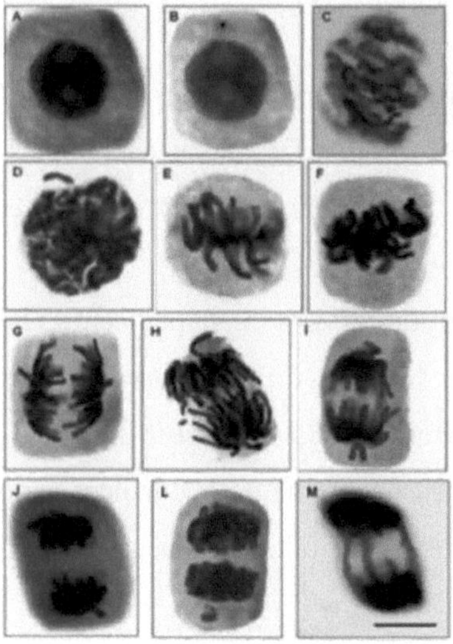

Figure 2. Mitotic cells obtained from different concentrations of T. grandis pyroligneous liquor. **A.** Normal interphase; **B.** Interphase with micronuclei; **C.** Normal prophase; **D.** Prophase with part of the chromosome isolated; **E.** Normal metaphase; **F.** Metaphase with chromosome isolated; **G.** Normal anaphase; **H** and **I.**

Anaphase with isolated chromosomes; **J.** Normal telophase; **L.** Telophase with isolated chromosome; **M.** Telophase with chromatin filament. Bar = 10 μM.

MACEDO et al. (2016) reported chromosomal abnormalities in A. cepa cells, finding the presence of fragmented chromosomes, chromatin bridges, lagging chromosomes and the presence of micronuclei in roots in contact with industrial wastewater. The results showed that there is cytogenetic alteration caused by pyroligneous liquor.

2.3 Infusion of Tabebuia avellanedae Lorentz ex. Griseb

The species Tabebuia avellanedae Lorentz ex. Griseb (Ipê-rosa), Bignoniaceae, occurs in the Atlantic Rainforest and is popularly known as: ipê-rosa-da-folha-larga, ipê-roxo-de-bola, ipê-rosado, ipê-de-minas, pau-d'arco, pau-d'arco-rosa, pau-d'arco-roxo, among others. The plant can reach a height of up to 30 m. Its leaves are compound, the inflorescence is a terminal panicle with numerous pinkish flowers, and the fruit is a cylindrical, elongated capsule with thin seeds with membranaceous wings (SOUZA et al., 2008).

The bark of T. avellanedae is used in folk medicine as an antifungal, diuretic and astringent tea; in the treatment of cancer, lupus, Parkinson's disease, psoriasis and allergies.

The results of its phytochemical analysis isolated the substances lapachol, lapachone, quercetin, hydroxybenzoic acid and some of their derivatives. Lapachol is easily extracted from wood sawdust as it is used for pharmaceutical purposes, which allows this material to be utilised (LORENZI et al., 2002).

There is a need to investigate the genotoxicity of T. avellanedae due to its widespread use in folk medicine. Based on this assumption, this study was carried out with the aim of verifying genotoxicity and its influence on the initial development and mitotic index of meristematic root cells of A. cepa.

Evaluations were carried out with three different concentrations: 9.75 g/200 mL (2 scoops) 16.08 g/200 mL (3 scoops), 17.88 g/200 mL (4 scoops), and control placed in a beaker (500 mL) and then 200 mL of water added at a temperature of 100 °C. The collected rootlets were washed and fixed in methanol: acetic acid (3:1) and stored under refrigeration at -4 °C for later use.

200 cells were analysed per treatment. The slides were prepared using the methodology of GUERRA et al. (2001). The mitotic index (MI) was obtained using the methodologies of OLIVEIRA et al. (1996) and PIRES et al. (2001). The means were subjected to analysis of variance and, for significant variations, the Tukey test was applied at 5% probability (FERREIRA, 2011).

The results obtained in the evaluations of the root behaviour of A. cepa from the control and the treatments with 9.75, 16.08 and 17.88 g of T. avellanedae showed a gradual reduction in the number of cells in mitosis. This was reflected in the MI, which was lower in the treatment with the highest concentration, 10.80 per cent in full cell division. As the concentrations increased, the number of cells in the interphase stage was also higher in relation to the other phases between the treatments (Table 3).

Table 3. A. cepa under different concentrations of T. avellanedae extract.

TREAT (g)	**NCM**	**Cells (%)**				**IM** (%)
		P	**M**	**A**	**T**	
0	1620	35	20,5	22,33	31,67	81
9,75	1240	61,88	0,17	0	0	62
16,08	220	8,33	1,67	0,17	0,33	11
17,88	217	8,33	1	0,83	0,67	10,8

Abbreviations: A = anaphase; MI = mitotic index; M = metaphase; NCM = number of cells in mitosis; P = prophase; T = telophase; TRAT = treatment.

The delay in the development of meristems or even the germination of seeds when in contact with allopathic substances will not necessarily cause the death of the plant, but will reduce its development by interfering with the formation of normal cells (CAMPOS, 2004).

At a concentration of 9.75 g of T. avellanedae there were no cells in the anaphase or telophase stage, while at a concentration of 16.08 g there were around 3.33 cells in anaphase and 6.67 in telophase (Table 4). There was an average increase in these two cell stages and a reduction in the number of cells in the metaphase phase. This may be associated with the action of genotoxic substances that may be interfering in the formation of the mitotic spindle and inhibiting development.

Among the different mitotic phases of A. cepa exposed to the three concentrations of ipê-rosa bark infusion (Table 4), there were significant differences between all the phases, but at concentrations of 16.08 g and 17.88 g the differences observed were less significant. This is probably due to the proximity of the concentrations and even the limit of genotoxic action on the root meristems.

Table 4: Averages of mitotic cells of A. cepa obtained by different treatments with T. avellaneda.

TREAT	**Mitotic division**				
	I	**P**	**M**	**A**	**T**

(g)					
0	380a	700b	410b	446,67b	63,33b
9,75	760b	1236,67c	3,33a	0a	0a
16,08	1780c	176,67a	33,33a	3,33a	6.67a
17,88	1783c	166,67a	20a	16,67a	13,33a
CV%	8,04	13,22	59,64	80,52	63,50

Abbreviations: A = anaphase; I = interphase; M = metaphase; P = prophase; T = telophase; TRAT = treatment.

Averages followed by different letters in the same column differ at the 5% probability level using the Tukey test.

During the analyses, in addition to the high number of cells in interphase (Figure 3A) and prophase (Figure 3B) that were observed in the treatments with the different solutions, cells were also observed in the process of cell death (Figure 3C). This process is probably associated with the chemical compounds released by T. avellanedae inhibiting cell activity.

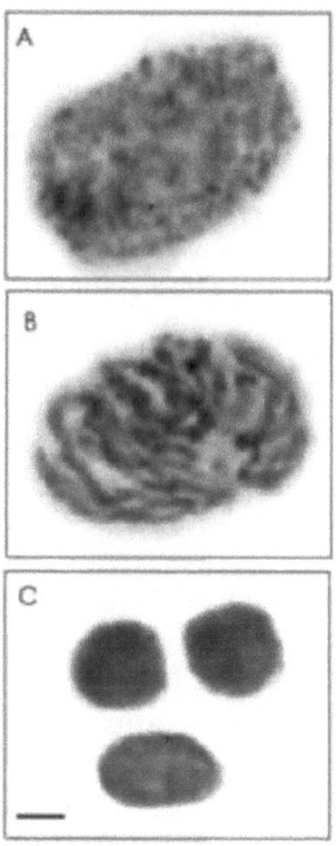

Figure 3: Cell division obtained from the A. cepa system in ipê-roxo infusion. **A.** Prophase; **B.** Interphase; **C.** Cells in the apoptosis stage. Bar = 10 ^M.

Chromosomal abnormalities were confirmed by GALVÃO et al. (2015), who showed cell death due to mutagenic potential in water collected at different points in the urban perimeter in the municipality of Alta Floresta using the A. cepa test. It is possible that the presence of these compounds in T. avellanedae is related to the inhibition of the A. cepa cell cycle observed in this study.

It is therefore concluded that the use of the A. cepa test system is a good alternative when evaluating the cytotoxic potential of ipê-rosa due to its simplicity and speed.

2.4 Infusion of the bark of the Hymenaea courbaril L. tree

Hymenaea courbaril L. is known in Brazil as jatobá (LORENZI, 2003). Belonging to the Fabaceae-Caesalpinioideae family, it occurs in almost all regions of Brazil. Most of the species in this genus are of some economic value, as well as having

The bark is used to prepare teas for flu, bronchitis, cystitis, diarrhoea, worms, weakness, colic, bladder infections, digestion and in the treatment of prostate cancer.

According to GADANO et al. (2002), many plants are useful medicinally, but more work needs to be done on toxicity (TEIXEIRA, 2014). The chromosomal abnormality procedure in A. cepa roots is an efficient evaluation process for diagnosis and monitoring, identifying the toxic effects and alterations that occur throughout a cell cycle (STURBELLE, 2010).

More studies are needed on the use of H. courbaril and its physiological, genotoxic and mutagenic effects on the human body (FIGUEIREDO, 2014). The aim of this study was to assess the sensitivity of biotests using A. cepa to monitor the phytocytotoxic effect of the medicinal species H. courbaril using the mitotic index.

Evaluations were carried out using aqueous extracts obtained by infusing the bark of H. courbaril at concentrations of 35, 65 and 130 g L^{-1} plus boiling water to obtain the extract for 15 minutes. The collected rootlets were washed and fixed in methanol: acetic acid (3:1) and stored under refrigeration at -4 °C for later use.

200 cells were analysed per treatment. The slides were prepared using the methodology of GUERRA et al. (2001). The mitotic index (MI) was obtained using the methodologies of OLIVEIRA et al. (1996) and PIRES et al. (2001). The means

were subjected to analysis of variance and, for significant variations, the Skott-Knott test was applied at 5% probability (FERREIRA, 2011).

In the results obtained from the evaluations of A. cepa germination in the aqueous extracts of H. courbaril, irregularities were found in the process of cell division in the different phases at the different concentrations (Figure 4).

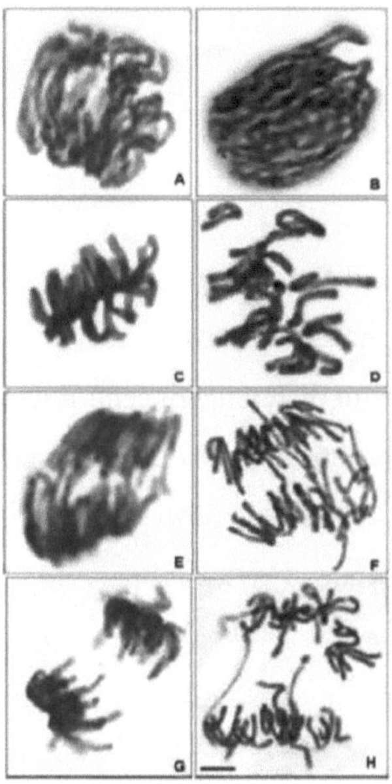

Figure 4: Mitotic stages of A. cepa obtained from the action of different concentrations of H. courbaril extract. **A.** Normal prophase; **B.** Prophase with isolated chromosome; **C.** Normal metaphase; **D.** Metaphase with isolated chromosomes; **E.** Normal anaphase; **F.** Anaphase with delayed chromosomes; **G.** Normal telophase; **H.** Telophase with bridge and chromosome outside the plate. Bar =10 gm.

In the prophase (Figure 4A) and metaphase (Figure 4D) stages, cells with isolated chromosomes were randomly visualised. In the anaphase (Figure 4F) and telophase (Figure 4H) stages, the formation of bridges and isolated chromosomes was found in all three treatments. There were significant differences in all the cell stages and different treatments and the highest number of cells in prophase was with the 35 g L^{-1} treatment (1710 cells).

The mitotic index, the percentage of cells that were in one of the four phases of mitosis, varied between 27.80% and 33.34% in the treatments analysed. The positive control had 100% of the cells analysed in full mitotic division. The mitotic division index differed between treatments, which may indicate a different physiological action of each of the extracts applied to the onion. The H. courbaril species can be considered genotoxic, as it showed irregularities in the cell division process in the different phases at the different concentrations.

2.5 Infusion and juice of Cecropia sp.

The Cecropia genus belongs to the Cecropiaceae family (LORENZI, 2002). The aim of this study was to analyse the genotoxic effect of different concentrations of tea and juice infusions from the Cecropia sp. genus using the A. cepa test.

Five treatments were used for the two treatments with different concentrations of green leaf tea and extracted from the juice of Cecropia sp. petioles, and one treatment with distilled water to characterise a control treatment. Meristematic tissues were extracted from the bulbs of each treatment and the mitotic index (MI) was analysed to count the cells. 2,500 cells were observed in each treatment.

The A. cepa test system showed that cell division was inhibited by all concentrations of Cecropia sp. infusion, except the control. These results prove the genotoxic action of Cecropia sp. as well as

prove the efficiency and safety of detecting toxic substances using the A. cepa test.

During the analyses, the total number of cells (2,500) and the number of cells in the different phases of the cell cycle (interphase, prophase, metaphase, anaphase and telophase) were assessed in the cells of A. cepa roots treated with Cecropia sp. using the green leaf and petiole juice. The cellular behaviour of the root meristems was also observed, indicating an increase in the number of cells in the interphase stage of mitosis as the concentration of the solutions increased, interfering with MI in cell division. This may influence the occurrence of irregularities in the different divisions due to the chemical substances that make up this extract.

Treatments with green leaf extracts of Cecropia sp. showed that increasing the concentrations of aqueous solutions of Cecropia sp. resulted in its inhibition, remaining in the interphase stage, not allowing cell division in meristematic cells of A. cepa to carry out the normal development process of cell division and consequently leading to cell death.

The mitotic index of A. cepa cells showed a significant decrease after 24 hours in relation to the treatments, at each concentration of Cecropia sp. extract, when compared to the control group with normal cells. It is possible that the presence of tannins in Cecropia sp. is related to the inhibition of the A. cepa cell cycle observed in this study.

Both analyses showed that the meristematic cells of A. cepa were inhibited in the process of cell division, similar to the work of DUSMAN et al. (2014). The mitotic index of the A. cepa cells showed a significant decrease in relation to the treatments compared to the control group, showing that Cecropia sp is highly cytotoxic.Figure 5 indicates the inhibition of the meristematic cells of A. cepa in the cell division process.

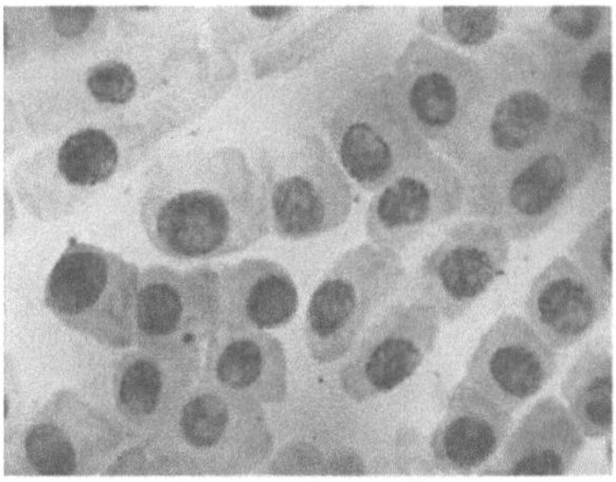

Figure 5: A. cepa interphases observed in different treatments with Cecropia sp.

Based on the results obtained and the conditions adopted, it can be mentioned that the evaluation of the consumption of Cecropia sp. analysed the inhibition of the meristematic cells of A. cepa in the process of cell division, which significantly reduced the mitotic index, indicating that it is highly cytotoxic.

The evaluation of Cecropia sp is in line with data from studies by BORGES et al. (2016) with the species Tradescantia pallida, which showed a significant percentage of sensitivity to exposure to contaminants. Therefore, studies on the compounds in medicinal plants are essential in order to obtain knowledge to prevent damage to public health.

2.6. Genotoxic effects and index of Tabebuia vellosoi Toledo bark infusions

Most of the medicinal plants used are not sufficiently studied in terms of their genotoxic potential. Given this situation, the aim of this study was to determine the genotoxic potential of Tabebuia vellosoi Toledo using the A. cepa test, since this species is widely used in folk medicine.

For the infusion, four different concentrations of peel were used, weighed on a precision scale: 0.5 tablespoon (1.177g), 1 tablespoon (1.914g), 1.5 tablespoons (3.617g) and 2 tablespoons (4.312g), as well as the control.

For the study, onion bulbs were exposed to four different concentrations for different lengths of time: 24, 48 and 72 hours, plus the control, where 150 cells per slide were analysed, totalling 1800 cells per treatment.

It was possible to observe that the greatest number of irregularities occurred in the 3.617 g treatment, emphasising that most of the irregularities occurred in the telophase stage. In the 1.177 g treatment, it was observed that, depending on the exposure time and concentration, the cells could have differentiated or even activated division peaks. As the exposure time of the A. cepa in the 1.914 g treatment increased, there was a decrease in the number of dividing cells, with the 24 and 48 h exposure times differing from 72 h (Figure 6). In the 4.312 g treatment, cell division remained constant. Given these results, it can be determined that the highest concentration of T. vellosoi is toxic, regardless of the exposure time (Figure 6).

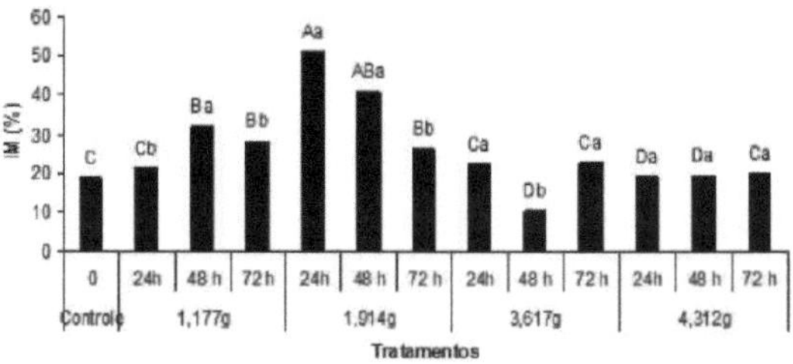

Figure 6: Mitotic division index of A. cepa cells treated with T. vellosoi bark extracts. Averages with different capital letters differ between treatments and lower case letters differ between treatments at the 5% probability level using the Scott-Knott test.

The A. cepa test showed that higher concentrations of ipê-amarelo bark resulted in lower cell division and an increase in the number of abnormal cells, indicating the toxicity of this species.

2.7. Infusions of Syzygium malaccense (L.) Merr. & L.M. Perry

Studies on the genotoxicity and mutagenicity of Syzygium malaccense are rare. With this in mind, this study was carried out with the aim of gathering information on its medicinal uses and applications and its genotoxicity and influence on cell division and the mitotic index of A. cepa root meristematic cells.

It was necessary to use and time exposures of green and dried S. malaccense leaf tea. The meristematic ends of the roots of the bulbs from each treatment were extracted and slides were made and 1500 cells per treatment were analysed. There was a significant difference between the control and the treatments. There was a reduction in the mitotic index between all the treatments compared to the control.

The S. malaccense leaves (12 g and 96 h) showed a lower mitotic index (0.25%) compared to the control test (8%), causing chromosomal alterations during cell division and inhibiting it (Figure 7), which shows that the infusion is a substance that inhibits other mitotic divisions.

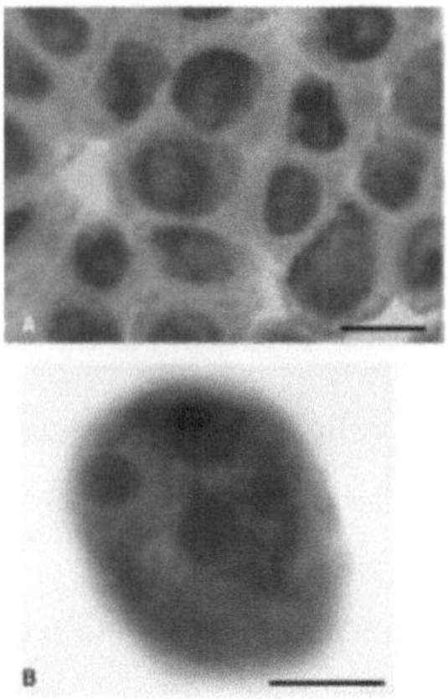

Figure 7. Mitotic cells of A. cepa obtained from exposure to S. malaccense infusion. **A.** Cells in prophase; **B.** Presence of micronuclei in cells in the prophase stage. Bar =10 pm.

Comparing the dry leaf and 6 g and dry leaf and 12 g treatments, a significant difference can be seen in the 24, 48 and 72 hour treatments. There was a decrease in the mitotic index, which was lower in the treatment with the (12 g) concentration. However, the same did not occur with the 96-hour treatment.

In the green leaf treatments (6 and 12 g), there was a decrease in the mitotic index as the concentration increased, visible in the 48 and 72 hour treatments. Analyses of meristematic cells from A. cepa bulbs showed that higher concentrations of extract reduced the mitotic index.

S. malaccense infusions have a genotoxic effect, since all concentrations cause cell cycle inhibition and a significant increase in the number of cells in interphase in the A. cepa test system. Further studies are needed to precisely define therapeutic dosages and periods of use, guaranteeing the safe use of this infusion, as well as maximising the benefits that S. malaccense can provide.

CHAPTER 3

Allium cepa TEST AS A BIOINDICATOR IN DIFFERENT CHEMICAL SUBSTANCES

3.1. Effect of the herbicide Diquat® on the cell cycle of Allium cepa

Agriculture, when considered in a broader sense, depending on the chemical nature and concentration of the pesticides released into the environment, can cause various types of damage to the biota exposed to them.

Although, in most cases, these compounds are not capable of causing acute and immediate effects, they can, on the other hand, reduce the survival of these organisms through chronic damage that manifests itself in the medium and long term as physiological disorders in different tissues and organs or as genetic alterations.

The aim of this study was to check the behaviour of mitotic cells in the A. cepa system after exposure to different concentrations of the herbicide Diquat® . Twenty A. cepa roots were used for each treatment. After the treatments, the root meristems were collected, washed in distilled water with three subsequent changes and fixed in methanol: acetic acid (3:1) until use. To make the slides, the meristems were washed with distilled water three times at 10-minute intervals.

Subsequently, the roots were hydrolysed with 1N HCl for 15 minutes and then washed in distilled water with two consecutive changes. The cells were stained with 2% acetic orcein.

4000 cells were evaluated for each treatment. In the cytotoxic evaluations carried out with A. cepa in the different treatments, it was found that in the negative control the cells showed all the mitotic phases, however in the other treatments the cells did not show the distinct phases, remaining almost all in the interphase phase where some cells carried out the mitotic cycle showing abnormalities such as metaphases with isolated chromosomes. The use of the herbicide caused cell death in both treatments

(Figure 8).

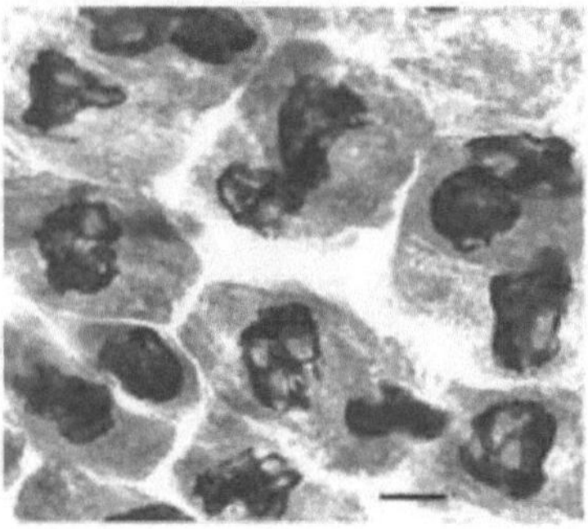

Figure 8. Mitotically dividing A. cepa cells exposed to different concentrations of herbicide. Bar = lOpim.

MACEDO et al. (2016) analysed the cell cycle of A. cepa after exposure to the herbicide Glyfosate® and observed that the herbicide caused a reduction in the mitotic division index, meaning that

so even at low concentrations the product causes changes in the cell cycle.

In the work by BEZERRA et al. (2016), genotoxicity was evidenced in the two highest concentrations analysed, as they showed a statistically significant frequency of chromosomal aberrations, corroborating the data from the present study, as the exposure of A. cepa root cells to the herbicide Diquat® showed that as the exposure time increased, the number of cells with irregularities grew, in addition to the observation of cells with apoptosis. This shows that you should carefully assess the crop you are working with and use a herbicide that doesn't affect the plant's development.

3.2. Allium cepa exposed to the pesticide Imazetapir®

The use of agrochemicals is the main strategy for combating animals and plants that harm the environment.

crops. The use of these chemicals both in production and in the home is becoming a concern, as it contributes to the release of toxic compounds into the environment.

Imazetapir® can be considered a possible contaminant of water and soil, as well as showing evidence of toxicological potential. It is a selective herbicide belonging to the imidazoline chemical group, designed to control broad-leaved weeds. Various studies have shown that some pesticides can be toxic to different organisms.

The aim of this study was to assess the mitotic behaviour of A. cepa exposed to a solution of the herbicide Imazetapir® at a concentration of 15%.

The treatments used were negative control (distilled water), positive control (acetone 15%) and herbicide (Imazetapir® 15%). Twenty onion roots were used for each sample of solution. The rootlets were exposed to the solutions for 72 hours, then removed and washed. To make the slides, the root meristem was crushed using the 2% acetic orcein dye

The 15% herbicide treatment showed a greater number of cells in the interphase stage, which differed from the other treatments. In addition, there was a significant decrease in the number of prophases between treatments and no metaphase, anaphase or telophase was seen (Figure 9), corroborating the data of MONTEIRO et al. (2016), who when assessing the behaviour of the A. cepa cell cycle in soil analysis containing domestic waste observed that the MI between the treatments used was increasing, indicating that soils with different contaminants showed different irregularities in mitosis.

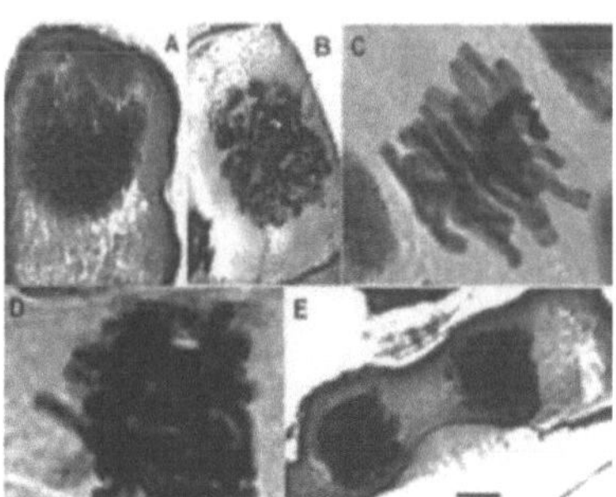

Figure 9. Mitotically dividing A. cepa cells exposed to different treatments with the herbicide Imazetapir® . **A.** Interphase; **B.** Prophase; **C.** Metaphase; **D.** Anaphase; **E.**

Telophase. Bar = 10μm.

The results of this study show that the pesticide analysed showed considerable variation in cell division, with both cytotoxic and antiproliferative effects being detected, in the same way that LIMA et al. (2016) with soil and coal subjected to the herbicide 2,4-dichlorophenoxyacetic acid (2,4-D) obtained variation between the genotoxic potential among the solutions tested.

3.3. Allium cepa exposed to the pesticide trifluralin®

Living organisms are frequently exposed to substances that can cause mutations in their cells. Damage can be induced by chemical, physical or biological agents that affect vital processes, leading to cancerous processes and cell death. The aim of this study was to verify the genotoxic effect of the herbicide trifluralin® in different concentrations (15% and 20%) using the A. cepa test system. Twenty A. cepa rootlets were used for each treatment.

To make the slides, the rootlets were washed with distilled water three times 10 minutes apart. They were then transferred to 1N HCl for 15 minutes and washed with distilled water for 10 minutes. 4000 cells were analysed per treatment.

Depending on the dosage applied and the period of exposure to the herbicide, anomalies were observed in the mitotic cycle and, in many cases, cell death. In the cells from the negative control treatment, the mitotic cell cycle was complete, with very distinct phases in total normality (Figures 10A, D, F-G).

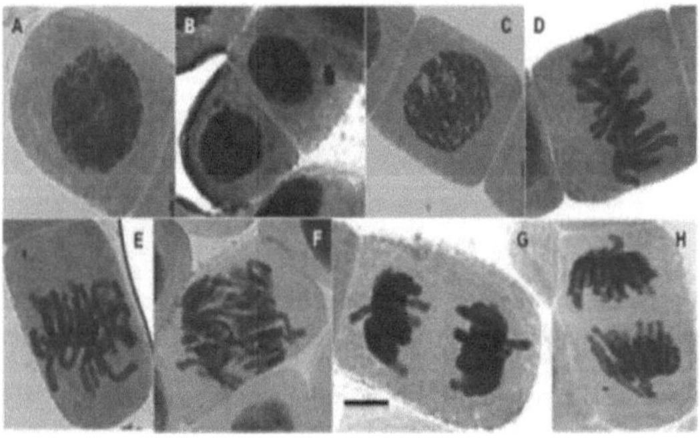

Figure 10. Mitotically dividing A. cepa cells exposed to different concentrations of the herbicide trifluralin® . **A.** Interphase; **B.** Interphase with micronucleus; **C.** Prophase; **D.** Metaphase; **E.** Metaphase with isolated chromatin; **F.** Anaphase; **G.** Telophase with lagging chromatids; **H.** Telophase with lagging chromatids and presence of isolated chromatid. Bar = 10μm.

When the A. cepa cells were subjected to the positive control (15% acetone) and the herbicide trifluralin® (15% and 20%), they behaved abnormally, showing a drop in the number of cells in the phases of the mitotic cycle and the presence of a micronucleus (Figures 10B, E and H).

Analysing the negative control A. cepa cells showed that they had a normal mitotic cycle, but when they were treated with the herbicide trifluralin® 15% and 20% they showed abnormalities and a decrease in cells in the last stages of mitosis. It is therefore possible to state that, depending on the dosage and time that the herbicide trifluralin® acts, it increases the irregularities in A. cepa cells, which can even lead to their death.

CHAPTER 4

EXPOSURE OF ALLIUM CEPA TO SOIL CONTAMINANTS

4.1. Cytogenotoxicity with distilled water mixed with soil sample

Pollution caused by the introduction of substances into the soil results in soil contamination. The alterations caused by these agents can generate mutations.

Bioindicators are biotic factors used to recognise ecosystem conditions by correlating a given anthropogenic factor or natural factor with the potential impact, representing an important tool in estimating ecological integrity (ANDRADE, 2010).

The use of the bioindicator A. cepa helps to assess the intensity of environmental pollution. Carrying out the test allows different concentrations to be observed, making it possible to assess the degree of soil contamination.

The aim of this study was to assess the cytotoxicity and genotoxicity of different soil samples using the A. cepa test system as a bioindicator, verifying the damage caused to root meristematic cells.

Two soil samples were collected and washed in equal proportions in distilled water. 25 onions were used for the control treatment. The concentrations soil sample 1, soil sample 2, charcoal exposed to the times of: 48, 72, 96 hours for each test system, finally 15 roots dipped in acetońe, subsequently removed, and 10 slides of each prepared for analysis. The collected rootlets were washed and fixed in methanol: acetic acid (3:1) and stored under refrigeration at -4 °C for later use. 2,500 cells were analysed per treatment. The slides were prepared following the methodology of GUERRA et al. (2001). The mitotic index (MI) was obtained using the methodologies of OLIVEIRA et al. (1996) and PIRES et al. (2001). The results obtained in the treatments of cells from the germination of A. cepa in the aqueous extracts of water with soil from two different locations showed irregularities.

The roots subjected to water containing sample 1 did not show any abnormalities during the cell division process at 48 hours. Unlike the results presented in the 72 and 96 hour treatments, in which abnormalities were detected in the interphase, prophase, anaphase, metaphase and telophase periods.

The cells analysed in sample 2 showed abnormalities in the 48-hour treatment for all periods of mitosis, but in the 72- and 96-hour treatments they showed few abnormalities only in the prophases, metaphases and telophases. In the charcoal treatments for 48, 72 and 96 hours, there were more irregularities in the prophases, metaphases, anaphases and telophases.

The mitotic index, the percentage of cells that were in the stages of mitosis, varied between 31.22% and 71.11% in the treatments analysed (Table 6). While the control had 100% of the cells analysed in full mitotic division. The mitotic division index differed between treatments, which may indicate a different physiological action of each of the extracts applied to the onion.

Table 6 Mitotic index of A. cepa obtained in different treatments with different soil samples and acetone.

Treatment	CN	CA	IM (%)
Witness	2500	0	31,22
1/48h	2250	0	32,49
1/72h	2495	45	36,10
1/96h	2478	22	43,36
2/48h	2482	18	43,30
2/72h	2491	9	71,11
2/96h	2493	7	69,17
Charcoal/48h	2433	67	33,40
Charcoal/ 72h	2491	9	51,72
Coal/96h	2471	29	33,66

Abbreviations: AC = abnormal cells; NC = normal cells; MI = mitotic index.

The data from the present study corroborate the information from DAMASIO et al. (2016), which showed an antiproliferative effect in A. cepa obtained from exposure to soils contaminated with domestic waste and did not show mutagenic potential when evaluated using the A. cepa bioassay.

It is concluded that A. cepa as a bioindicator showed irregularities in the stages of mitosis, as did SOARES et al. (2016) using mitotic changes in A. cepa with soil treatments, as it showed irregularities in the process of cell division in the different phases at the different concentrations.

The mitotic index showed differences between treatments and the appearance of these abnormalities shows that the cytotoxicity induced by the substances present in the soil samples potentially interferes with the cell cycle.

CHAPTER 5

MUTAGENESIS THROUGH THE BIOINDICATOR Lactuva sativa

The species Lactuca sativa L. belongs to the Asteraceae family and originated in Europe. It is sensitive to environmental conditions and can cause problems with germination and emergence. They also have various medicinal properties, such as: antacid, antirheumatic, calming of the stomach and nervous system (TEIXEIRA, 2014).

L. sativa is used to assess toxicity and environmental pollution by measuring the percentage of germination and root elongation. Root growth is one of the most sensitive parameters to a phytotoxic agent and is affected by environmental conditions such as pH, temperature, salinity and the presence of metals.

5.1 Exposure of Lactuva sativa to herbal infusions, juices and macerates

5.1.1 Mutagenesis of the aqueous solution of Hymenaea courbaril L. var.stilbocarpa

Evaluating the sensitivity of biotests using L. sativa seeds to monitor the phytocytotoxic effect of Hymenaea Courbaril L. var. stilbocarpa using the mitotic index (MI), the presence of chromosomal alterations was observed, such as isolated chromosomes (Figures 11B and D), chromatid breaks, telophase bridges (Figures 11F and H) and cell death (Figure 11 I).

Similarly to A. cepa, L. sativa cells showed a greater number of irregularities in the anaphase and telophase stages and cell death occurred, an activity that jeopardises the development of root meristems and the plant in general.

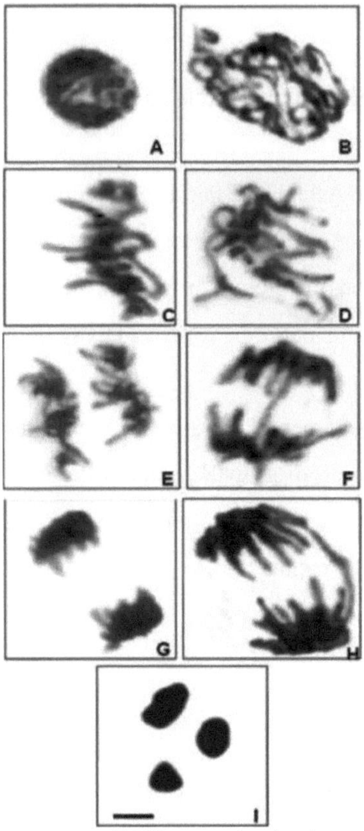

Figure 11. Different mitotic phases of L. sativa obtained from the action of different concentrations of the extract

of H. courbaril. A. Normal prophase; B. Prophase with isolated chromosomes; C. Normal metaphase; D. Metaphase with isolated chromosomes; E. Normal anaphase; F. Telophase with bridge in median portion; G. Normal telophase; H. Telophase with distal bridge; I. Cell death. Bar =10 μm.

According to CUCHIARA (2012), as the concentration of plant extracts rich in toxic substances such as alkaloids and tannins increases, the tendency is to reduce the number of dividing cells, increasing the chances of cell death.

L. sativa seeds showed the highest germination rate on the second day after exposure to the different Hymenaea courbaril extracts, at a concentration of 65g. The mitotic index differed significantly between the positive control and the L. sativa bioincinerator. The Hymenaea courbaril species can be considered genotoxic, as it showed irregularities in the cell division process in the different phases at the different concentrations.

4.1.1. Cytotoxicity using Morinda citrifolia fruit extract in natura

The behaviour of L. sativa cells was assessed in different Morinda citrifolia solutions and with different exposure times, monitoring potentially toxic substances. To assess cytotoxicity, the extract of the fruit of M. citrifolia in natura and the test organisms with L. sativa seeds were used. Concentrations of 5, 10, 15 and 20 grams diluted in 200 mL of distilled water were used for each assessment, a negative control treatment with distilled water only and a positive control using 15% acetone.

The treatments of L. sativa using the aqueous extracts and M. citrifolia showed that L. sativa had a MI of 3.48% for 5 g and 2.93% for 10 g after 24 hours of exposure.

For L. sativa, cells with abnormalities were observed in the negative and positive controls. The negative control showed an average of 0.05% and the positive control 0.18%, while the treatment with the highest average among the concentrations of M. citrifolia extract showed no abnormalities. The L. sativa test system proved to be less sensitive to exposure to the aqueous extract of M. citrifolia than the A. cepa biomarker.

4.1.2. Allelopathy of aqueous extracts of Annona muricata L.

As well as being a fruit species with strong economic potential, Annona muricata also has medicinal properties and is widely used to treat diarrhoea and spasms, as well as some types of cancer (LORENZI & MATOS, 2002).

Most allelopathic studies are restricted to invasive plants and evaluations of fruit trees with economic and medicinal value, such as A. muricata, for example, are still scarce. Bearing in mind that this species could be an important source for the production of chemical and biological compounds, this study aimed to evaluate the allelopathic potential of the aqueous extract of graviola leaves on the germination and initial development of L. sativa.

Two aqueous extracts were obtained by infusing and macerating the leaves. The concentrations used were: 0; 0.05; 0.10; 0.20; and 0.40 mg L^{-1} . The bioassay consisted of five treatments with four replicates of 50 seeds. The following characteristics were assessed: first germination count, germination speed index, germination percentage, area and root length and fresh and dry mass of the seedlings.

The experiment assessed the first count at four days, the germination test at seven days and the germination speed index, in which the number of germinated seeds was counted daily.

To assess the development of the seedlings, fifty seedlings per treatment were chosen at random and the length of the root part and the aerial part were measured using a digital caliper. The fresh and dry mass of the seedlings was estimated using a precision scale.

The data was submitted to analysis of variance and for significant variations, the Tukey test was used at 5% probability using the Sisvar programme (FERREIRA, 2011).

The results obtained from the two extracts (macerate and infusion) showed that there was a greater allelopathic effect for the infusion treatment, with less root development. In relation to the germination of lettuce seeds, it was found that the concentration with the greatest allelopathic effect was 0.40 mg, showing the greatest inhibition in the germination speed index. The concentration with the lowest allelopathic effect was 0.05 mg.

The different concentrations of the aqueous extracts of graviola leaves acted differently; the higher the concentration of extract, the greater the allelopathic effect and the lower the germination speed index. These results corroborate those of MACEDO et al. (2016), who assessed the allelopathic potential of different treatments and concentrations of Melissa officinalis leaf on the germination and initial development of L. sativa and concluded that the treatment using macerated dry leaf at concentrations of 0.20 g L^{-1} and 0.10 g L^{-1} showed a lower germination index and lower average root part, thus indicating a greater allelopathic potential.

CHAPTER 6

use of the bioindicator pisum sativum linnaeus in mutagenesis

The Pisum sativum Linnaeus pea belongs to the order Fabales, Family Fabaceae and the genus Pisum. The pea is a vegetable with a great deal of variation, easy to digest and nutritious. It is a good source of minerals such as phosphorus, calcium, potassium, iron and vitamins A and C.

When it comes to health, the benefits of peas are for the heart. It also helps prevent plaque formation in the arteries, reduce LDL cholesterol and triglyceride levels, diabetes and prevent cancer.

Among the test systems, the P. sativum test system can be used to study the effects of plant extracts in order to detect genotoxicity, but little is known about this bioindicator.

6.1. Evaluation of the genotoxic effects of Tectona grandis Linn. f

Tectona grandis Linn. f, popularly known as teak, is a large, fast-growing tree species with noble wood of excellent quality and is native to south-west Asia. It is a species that is expanding in the state of Mato Grosso.

The aim of this study was to assess the genotoxic effects of T. grandis on P. sativum, checking its mitotic behaviour at different exposure times.

The test organisms used were pea seeds which were sown in petri dishes. Ten P. sativum seeds were used for each treatment. After rooting, they were treated with 4 mL L^{-1} of pyroligneous liquor obtained from the combustion of teak wood exposed for 24, 48, 72 and 96 hours, during which time the liquor was changed daily. The collected rootlets were washed and fixed in methanol:acetic acid (3:1) and stored under refrigeration at -4 °C for later use. 300 cells were analysed per treatment.

The slides were prepared using the methodology of GUERRA et al. (2001). The mitotic index (MI) was obtained using the methodologies of OLIVEIRA et al. (1996) and PIRES et al. (2001). The means were subjected to analysis of variance and for significant variations the Tukey test was applied at 5% probability (FERREIRA, 2011).

The results of the treatments showed an increasing number of cells in the interphase phase. The percentage in the 24, 48, 72 and 96 hour treatments was higher than that found in the control treatment, ranging from 297.6% to 300% in ascending order.

Cells with abnormalities could be seen in the 24, 48 and 72 hour treatments, with abnormal metaphases and abnormal prophases being visualised. The highest percentage of abnormal cells occurred in the 48-hour treatment, with 1.3% of abnormal metaphases (Figure 12).

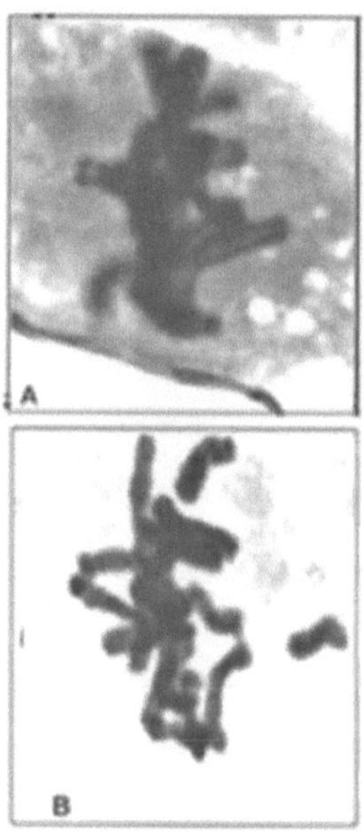

Figure 12. Mitotic metaphase of P. sativum, showing normal metaphase (**A**) and abnormal metaphase with isolated chromosome (**B**).

The teak extract at a concentration of 4 mL L^{-1} has a genotoxic effect, as it can be seen that over the course of the exposure time it caused cell cycle inhibition in the P. sativum bioindicator. The P. sativum test system proved to be a good alternative for assessing genotoxic potential due to its sensitivity to the allelopathic effects of the extract analysed, thus showing that this bioindicator can be used to monitor the action of toxic substances.

6.2. Genotoxicity of Centella asiatica L. Urb.

Centella asiatica is a plant native to Brazil, which has been used to treat dementia, asthma, ulcers and eczema and to heal wounds. However, there is little data on the possible health risks of prolonged use of medicinal plants.

One of the ways to identify these possible harmful effects is through bioindicators. Although little is known about the P. sativum test system, this bioindicator, like A. cepa, can be used to study the effects of plant extracts in order to detect genotoxicity (MELLO et al., 2015). The aim of this study was to assess the genotoxicity of P. sativum through mitotic behaviour at different exposure times to the aqueous extract of C. asiatica.

The test organisms used were P. sativum seeds. The seeds were placed in Petri dishes, lined with filter paper and moistened with distilled water. The Petri dishes were placed in a germination chamber. Twenty-five P. sativum seeds were used for each treatment. After rooting, they were treated with 0.93, 1.86 and 3.72 g L^{-1} of processed C. asiatica leaves purchased from a commercial establishment, infused for 15 minutes at approximately 100 °C and used at room temperature.

The rootlets were exposed for 24, 48, 72, 96 and 120 hours. The control treatment was also carried out and left in distilled water for 24 hours. After the treatments, the root meristems were collected, then washed in distilled water, fixed in methanol:acetic acid (3:1) and stored in the refrigerator.

The slides were prepared using the methodology of GUERRA et al. (2001). The mitotic index (MI) was obtained using the methodologies of OLIVEIRA et al. (1996) and PIRES et al. (2001). The means were subjected to analysis of variance and for significant variations the Tukey test was applied at 5% probability (FERREIRA, 2011).

In all the treatments analysed, the mitotic index with the highest percentage of cells in one of the four phases of mitosis occurred in the treatment 1.86 g L^{-1} and 120 hours with 48.16 % of cells in division, followed by 23.48% in the treatment 3.72 g L- and 120 hours. For the AP variable, treatment 1.86 g L^{-1} and 24 hours showed the highest percentage of abnormal cells at 2.52% (Table 6).

Table 6: Mitotic index (MI) and percentage of anomalies (PA) of P. sativum in different treatments.

Treatment	IM	PA
0 g L-1 and 24 h	4,60 c	0,00 c
0.93 g L^{-1} and 24 h	2,16 c	1.16 abc
0.93 g L^{-1} and 48 h	0,20 c	0,00 c
0.93 g L^{-1} and 72 h	7,08 c	2.28 ab
0.93 g L^{-1} and 96 h	5,40 c	0,16 c
0.93 g L^{-1} and 120 h	18,08 b	1.04 abc
1.86 g L^{-1} and 24 h	4,12 c	2,52 a
1.86 g L^{-1} and 48 h	5,84 c	0,40 c
1.86 g L^{-1} and 72 h	4,36 c	1.56 abc
1.86 g L^{-1} and 96 h	3,80 c	0.60 bc
1.86 g L^{-1} and 120 h	48,16 a	1.44 abc
3.72 g L^{-1} and 24 h	2,60 c	0,32 c
3.72 g L^{-1} and 48 h	5,20 c	0,00 c
3.72 g L^{-1} and 72 h	5,68 c	2.24 ab
3.72 g L^{-1} and 96 h	4,12 c	0.60 bc
3.72 g L-1 and 120 h	23,48b	1.60 abc
CV(%)	52,78	75,73

Averages followed by different letters in the same column differ at the 5% probability level using the Tukey test.

Looking at the low mitotic index (MI), it can be said that this is due to the high number of cells in interphase. It can also be said that the C. asiatica extract caused a genotoxic effect due to the occurrence of cells with anomalies and that with increasing concentrations there was an increase in cell division and an increase in chromosomal aberrations.

CHAPTER 7

FINAL WORDS

The results obtained in these studies, through the genotoxicity and mutagenicity tests carried out with the bioindicators A. cepa, L. sativa and P. sativum, showed that most species were sensitive to the effects caused by exposure to the aqueous extracts. The methodologies used in the studies were the same, proving to be efficient in the work carried out at the Cytogenetics and Plant Tissue Culture Laboratory of the State University of Mato Grosso at the Alta Floresta Campus - MT, even though each study required adjustments to the treatments and species studied.

CHAPTER 8

REFERENCES CITED

ALVES, L. A.; BARRETO, I. C; BORGES, J. M. P. Evaluation of cytotoxicity and genotoxicity induced by Stryphnodendron barbatimam on the cell cycle of Allium cepa. In: **Anais**... SCIENTIFIC INITIATION WEEK, 5, EXTENSION SYMPOSIUM, 1. Sinop, 2013.

ANDRADE, B. R. G. **Study of the cytotoxic, genotoxic and mutagenic potential in** Allium cepa **cells of** Parathion methyl **before and after application of the UV and** UV/H2O2 **processes**. Thesis (Doctorate in Chemical and Biochemical Process Technology, School of Chemistry) - Federal University of Rio de Janeiro. Rio de Janeiro, 2012.

ANDRADE, H. A. **Global scientometric analysis of bioindicators: an overview of trends between 1998 and 2007**. Dissertation (Master's Degree in Environmental Sciences and Health) - Pontifical Catholic University of Goiás. Goiânia, 2010.

BEZERRA, C. M.; OLIVEIRA, M. A. S. Evaluation of the toxicity, cytotoxicity and genotoxicity of mallow infusion (Plectranthus barbatus - Lamiaceae) on the cell cycle of Allium cepa. **Electronic Journal of Pharmacy**, 13(4): 220-228, 2013.

BORGES, N. M.; CARDOSO, M. A.; SOARES, J. A.G; WONS, D. M.; MELLO,V.; KARSBURG, I. V. Pollen viability and genotoxicity in Tradescantia pallida (Rose) D.R Hunt var. Purpurea bom (Commelinaceae) in the city of Alta Floresta-MT. **Scientific Electronic Archives: Special Edition**, 8486 2016.

BRITO, D. R. B. **Evaluation of the anti-helminthic activity of** Morinda citrifolia **(noni) in naturally infected laying birds**. Thesis (Doctorate in Animal Health and Reproduction) - Federal University of Piauí. Teresina, 2008.

CAMPOS, J. M. S. **Cytogenetic investigation of the allelopathic potential of lichen and pteridophyte species**. Dissertation (Master's Degree in Genetics and Plant Breeding) - Federal University of Lavras, 2004. 97p.

CARITA, R.; MARIN-MORALES, M. A. Induction of chromosome aberrations in the Allium cepa test system caused by the exposure of seeds to industrial effluents contaminated with zo dyes. **Chemosphere**, 72(5): 722725, 2008.

CUCHIARA, C. C.; BORGES, C. S.; BOBROWSKI, V. L. Allium cepa test system as a bioindicator of cytogenotoxicity of watercourses. **Tecnologia. Ciência Agropecuária.** 6(1): 33-38, 2012.

DAMASIO, J. F.; KARSBURG, I. V.; MONTEIRO, M.; LIMA, E. S. Evaluation of the meiotic index of a population of Psychotria ipecacuanha (Rubiaceae) from Nova Olímpia, Mato Grosso, Brazil. **Scientific Electronic Archives: Special Edition, 84-86.** 140142.,2016.

DAMASIO, J. F.; KARSBURG, I. V.; SANTOS, J. F.; LAZAROTTO, L.; LUCA, K.; HENRIQUE, M. Antiproliferative effect on Allium cepa obtained from exposure to soils contaminated with domestic waste.
Scientific Electronic Archives: Special Edition, 137-139. 2016.

DA SILVA, A. E. P.; MIRANDA, J. W.; NETO, M. P. L. Toxic, cytotoxic, genotoxic and mutagenic evaluation of Turnera ulmifolia L. (Chanana) in eukaryotic cells. **Health in Focus**. 2(1): 25-48, 2015.

DUSMAN, *E;* LUZZA, M.; SAVEGNAGO, L.; LAUXEN, D.; VICENTINI, V. E.; TONIAL, I. B.; SAUER, T. P. Allium cepa L. as a bioindicator to measure cytotoxicity of surface water of the Quatorze River, located in Francisco Beltrão, Paraná, Brazil. **Environmental Monitoring and Assessment**. 186(3): 1793-800, 2014.

FACHINETTO, J. M.; BAGATINI, M. D.; DURIGON J.; SILVA, A. C. F.; TEDESCO, S. B. Anti-proliferative effect of infusions of Achyrocline satureioides DC (Asteraceae) on the cell cycle of Allium cepa. Revista Brasileira de Farmacognosia 17(1): 49-54, 2007.

FACHINETTO, J. M.; TEDESCO, S. B. Antiproliferative and mutagenic activity of aqueous extracts of Baccharis trimera (Less.) A. P. Candolle and Baccharis articulata (Lam.) Pers. (Asteraceae) on the Allium cepa test system. **Brazilian Journal of Medicinal Plants** 4(11): 360-367, 2009.

FERNANDES, F. L.; SANTOS, L. C. B.; MACEDO. W. A.; LEITE, D. M.; DAMASIO, J. F.; KARSBURG, I.V. Effect of the aqueous extract of Centella asiatica L. Urb. (Apiaceae) on the mitotic index of Allium cepa L. and Pisum sativum L.

Scientific Electronic Archives: Special Edition, 161-163. 2016.

FERNANDES, T. C. C.; MAZZEO, D. E. C.; MARIN- MORALLES, M. A. Mechanism of micronuclei formation in polyploidised cells of Allium cepa exposed to trifluralin herbicide. **Pesticide Biochemistry and Physiology**, 88, 252-25, 2007.

FERREIRA, D. F. Sisvar: a computer statistical analysis system. **Ciência e Agrotecnologia**. 35(6): 1039-1042, 2011.

FIGUEIREDO, D. R. **Evaluation of the cytotoxicity of the hydric extract of fennel** (Pimpinella anisum **L.) through the test on** Allium cepa **L.** Final Course Work (Undergraduate Degree in Biological Sciences) - State University of Paraíba. Campina Grande, 2014. 19f.

FISKESJO, G. The Allium test. **Environmental Toxicology and Water Quality**. 8: 291-298, 1993. GADANO, A.; GURNI, A.; LÓPEZ, P.; FERRARO, G.; CARBALLO, M. In vitro genotoxic evaluation of the medicinal plant Chenopodium ambrosioides L. **Journal of Ethnopharmacology** 81: 11-16: 2002.

GADELHA, C. S.; JUNIOR, V. M. P.; BEZERRA, K. K. S.; PEREIRA, B. B. M.; MARACAJÁ, P. B. Bibliographical study on the use of medicinal plants and herbal medicines in Brazil. **Green Journal of Agroecology and Sustainable Development**. 8(5): 208-212, 2013.

GALVÃO, M. Genotoxicity index by Allium cepa in the bioindicator test in waters polluted by chemical waste in the municipality of Alta Floresta - MT **Enciclopédia Biosfera**. 11(21): 2373-2383, 2015.

GUERRA, M.; SOUZA, M. J. **How to observe chromosomes - a guide to techniques in plant, animal and human cytology**. FUNPEC: Ribeirão Preto, 2001.

HEINICKE, R. M. **The pharmacologically active ingredient of Noni**. Bulletin of the National Tropical Botanical Garden, 1985.

KNOLL, M. F.; SILVA, A. C. F.; CANTODOROW, T. S.; TEDESCO, S. B. Effects of Pterocaulon polystachyum DC. (Asteraceae) on onion (Allium cepa) root-tip cells. **Genetics and Molecular Biology -**. 29: 539-542, 2006.

LEITE, D. M.; SANTOS, E.; SANTOS. J. F.; FERNADA, J.; TAKAHASI, M.; KARSBURG, I. V. Effect of the herbicide Diquat® on the cell cycle of Allium cepa

(Liliaceae). **Scientific Electronic Archives: Special Edition**, 253-255. 2016.

LEME, D. M.; ANGELIS, D. F.; MARIN-MORALES, M. A. Action petroleum hydrocarbons present in water impacted by an oil spill on the genetic material of Allium cepa root cells. **Aquatic toxicology**. 88(n): 214219, 2008.

LIMA, E. S.; MACEDO, W. A.; SANTOS, B. N. V.; CARDOSO, M. A.; DAMASIO, J. F.; KARSBURG, I. V. Allelopathy of aqueous extracts of Annona muricata L. (Annonaceae) on the germination and initial growth of Lactuca sativa L. (Asteraceae). **Scientific Electronic Archives: Special Edition**, 264-266. 2016.

LORENZI, H. **Árvores brasileiras: manual de identificação e cultivo de plantas arbóreas nativas do Brasil.** 3ª Ed. Nova Odessa: Instituto Plantarum, 2000. 46p.

LORENZI, H. **Exotic trees in Brazil: timber, ornamental and aromatic**. Nova Odessa: Instituto Plantarum, 2003. 384p.

LORENZI, H.; MATOS, F. J. A. **Plantas medicinais no Brasil: nativas e exóticas.** Nova Odessa: Instituto Plantarum, 2002. 93p.

MACEDO, W. A.; ALVES, V. C. R.; CRUZ, V. S.; SANTOS, B. N. V.; KARSBURG, I. V. Cell cycle of Allium cepa L. (Liliaceae) after exposure to the herbicide Glyfosate. **Scientific Electronic Archives: Special Edition,** 279-281. 2016.

MACEDO, W. A.; SANTOS, B. N. V.; LIMA, E. S.; SANTOS, L. C. B.; MELLO, V. S.; CARDOSO, M. A.; KARSBURG, I. V. Allelopathic potential of Melissa officinalis L. (Lamiaceae) leaves on lettuce germination. **Scientific Electronic Archives: Special Edition,** 282-284. 2016.

MASCHIO, L. R. **Evaluation of the cytotoxic, genotoxic and mutagenic potential of the waters of the Rio Preto in the area of influence of the region of São José do Rio Preto/SP**. Thesis (Doctorate in Genetics) - Paulista State University. São José do Rio Preto, 2009. 208p.

MENDES, M. S. Trace elements in Allium cepa L. and Lactuca sativa L. Postgraduate Programme in Agronomy. Federal University of Pelotas 2008.

MELLO, V. D. S.; MIRANDA, D. P.; DA SILVA, D. D.; MACHADO, D.; DA SILVA, A. B.; DAHMER, N.; KARSBURG, I. V. Genotoxic effect of teak pyroligneous liquor on the pea bioindicator. **Studies.** 41, 141-146, 2015.

MONTEIRO, M.; KARSBURG, I. V.; DAMASIO, J. D.; FRACARO, R. B.; SILVA, P. A.; SOUSA, R. L.; Behaviour of the cell cycle of Allium cepa in soil analysis containing domestic waste. **Scientific Electronic Archives: Special Edition,** 299-301. 2016.

MOROSINI; R. R.; KARSBURG. I. V. **Cytotoxic evaluation of aqueous extracts of** Tabebuia avellanedae **Lorentz ex Griseb (Ipê-Rosa)**. Final Course Work (Bachelor's Degree in Forestry Engineering) - Mato Grosso State University. Alta Floresta, 2008. 32p.

MOURA· R. T.; KARSBURG, I. V. **Cytogenotoxicity of** Tectona grandis **pyroligneous liquor by** Allium cepa. Final Course Work (Bachelor's Degree in Forestry Engineering) - Mato Grosso State University. Alta Floresta, 2013. 10p.

MULLER, J. C. **Reproductive toxicity of *Morinda citrifolia* Linn**. Dissertation (Master's Degree in Pharmacology) - Federal University of Paraná. Curitiba, 2007. 103p.

NATARAJAN, A. T. Chromosome aberration: past, present and future. **Mutation Research**. 504(6): 3-16, 2002.

OBE, G.; PFEIFFER, P.; SAVAGE, J. R. K.; JOHANNES, C.; GOEDECKE, W.; JEPPESEN, P.; NATARAJAN, A. T.; MARTINEZ-LÓPEZ, W.; FOLLE, G. A.; DRETS, M. E. Chromosomal aberrations: formation, identification and distribution. **Mutation Research**. 504(5), 17-36, 2004.

OLIVEIRA. E. M. **Evaluation of the cytotoxicity of *Morinda citrifolia* (Linn) *in natura*.** Course Conclusion Work (Full Degree in Biological Sciences) - Mato Grosso State University. Alta Floresta, 2013. 36p.

OLIVEIRA, J. M. Evaluation of the mutagenic potential of the waters of the tributaries of the Boa Vista River in the municipality of Ouro Preto do Oeste-RO.Dissertation. 2014.

OLIVEIRA, V. R.; SCAPIM, C. A.; OLIVEIRA JÚNIOR, R. R. S.; PIRES, N. M. Effect of the herbicide trifluralin on seed germination and mitotic index in maize (Zea mays L.) roots. **Revista Unimar**. 18(1): 537-544, 1996.

PIRES, N. M.; SOUZA, I. R. P.; PRATES, H. T.; FARIA, T. C. L.; FILHO, I. A. P.; MAGALHÃES, P. C. Effect of leucena aqueous extract on development, mitotic index and peroxidase activity in maize seedlings. **Revista Brasileira de Fisiologia Vegetal.** 13(1): 55-65, 2001.

SANTOS, B. N. V.; VIANA, T. L.; SANTOS, A. C.; MIRANDA, D. P.; MACEDO, W. A.; KARSBURG, I. V. Genotoxic potential and antiproliferative effect on the cell cycle of Allium cepa L. (Liliaceae) exposed to pesticides. **Scientific Electronic Archives: Special Edition,** 420-422. 2016.

SANTOS, R. L.; GUIMARÃES, G. P.; NOBRE, M. S. D. C.; PORTELA, A. D. S. Analysis of phytotherapy as an integrative practice in the Unified Health System. **Brazilian Journal of Medicinal Plants**. 13(4): 486491, 2011.

SILVA, A.; JACOBI, F. T.; SILVA, D. D.; MAIA, R. V.; KARSBURG, I. V. Genotoxicity of the herbicide trifluralin by the Allium cepa system. **Scientific Electronic Archives: Special Edition,** 435-437. 2016.

SILVA, C. G. **The cytotoxicity of Barbatimão (**Stryphnodedron adstringens **(Mart) Coville) using the** Allium cepa **and** Psisum sativum **systems**. Final Coursework (Full Degree in Biological Sciences) - Mato Grosso State University. Alta Floresta, 2012.

SOARES, J. A. G.; LEITE, D. M.; BORGES, N. M.; FERREIRA, F. S.; XISTO, L.; KARSBURG, I. V. Environmental genotoxicity using mitotic changes in Allium cepa (Liliaceae) as a bioindicator.

Scientific Electronic Archives: Special Edition, 465-467. 2016.

SOUZA, L. A.; OLIVEIRA, J. H. G. Morphology and anatomy of Tabebuia avellanedae Lor. ex Griseb and T. chrysotricha (Mart. ex Dc.) Standl.(Bignoniaceae) seedlings. **Acta Scientiarum. Biological Sciences**. 26(2): 217-226, 2008.

SOUZA, S. A. M.; STEIN, V. C.; CATTELAN, L. V.; BOBROWSKI, V. L.; ROCHA, B. H. G. Use of lettuce and rocket seeds as biological tests to evaluate the cytotoxic and allelopathic effects of aqueous extracts of medicinal plants. **Journal of Biology and Earth Sciences**. 5(1): 3-9, 2005.

SOLANO, J. D; SOARES, J. A. G.; BORGES, N. M.; ARAUJO, D. S.; KARSBURG, I. V. Evaluation of the mutagenic effect of the aqueous solution of Hymenea coubaril L. using the Allium cepa and Lactuca sativa biotests. **Enciclopédia Biosfera**. 11(21): 314324, 2015.

STURBELLE, R. T.; PINHO, D. S. D.; RESTANI, R. G.; OLIVEIRA, G. R. D.; GARCIAS, G. D. L.;

MARTINOROTH, M. D. G. Evaluation of the mutagenic and antimutagenic activity of Aloe vera in the Allium cepa test and micronucleus test in binucleated human lymphocytes. **Revista Brasileira de Farmacognosia**. 20(3): 409-415, 2010.

TEIXEIRA, S.A.; MELO, J. I. M. Medicinal plants used in the municipality of Jupi, Pernambuco, Brazil. **Iheringia Série Botânica**. 61(1): 05-11, 2014.

TRAESEL, G. K.; CASTRO, L. H. A.; SILVA, P. V. B.; MUZZI, R. M.; KASSUYA, C. L.; ARENA, A. C.; PIETA, S. A. Assessment of the cytotoxic, genotoxic, and mutagenic potential of Acrocomia aculeata in rats. **Genetics and Molecular Research**. 14(1): 585-596, 2015.

WANG, M. Y., WEST, B. J., JENSEN, C. J., NOWICKI, D., SU, C., PALU, A. K., & ANDERSON, G. Morinda citrifolia (Noni): A literature review and recent advancesin Noni research. **Acta Pharmacologica Sinica**. 23(12): 1127-1141, 2002.

More Books!

yes

I want morebooks!

Buy your books fast and straightforward online - at one of world's fastest growing online book stores! Environmentally sound due to Print-on-Demand technologies.

Buy your books online at
www.morebooks.shop

Kaufen Sie Ihre Bücher schnell und unkompliziert online – auf einer der am schnellsten wachsenden Buchhandelsplattformen weltweit! Dank Print-On-Demand umwelt- und ressourcenschonend produzi ert.

Bücher schneller online kaufen
www.morebooks.shop

info@omniscriptum.com
www.omniscriptum.com

OMNIScriptum

MIX
Papier aus verantwortungsvollen Quellen
Paper from responsible sources
FSC® C105338

Printed by Books on Demand GmbH, Norderstedt / Germany